Virginia Living Museum

Wild about Nature!

Great Horned Owl

Karl von Rebenstorff

Virginia Living Museum
Wild about Nature!

Edited by Page Hayhurst

Dedicated to the Virginia Living Museum Family

Conservation Notes

The Virginia Living Museum connects people to nature through educational experiences that promote conservation. Look for this symbol to learn more about our conservation work.

Cover Photo of Bobcat by Karlton Rebenstorf
Back Cover Photo by Sara Harris Photography

Published by The Donning Company Publishers
184 Business Park Drive, Suite 206, Virginia Beach, VA 23462

Virginia Living Museum
524 J. Clyde Morris Boulevard
Newport News, VA 23601

The Virginia Living Museum is a private, non-profit organization made possible by the generous support of donors and the in-kind service of volunteers.

Library of Congress Cataloging-in-Publication Data

Virginia Living Museum : wild about nature! / edited by Page Hayhurst.
p. cm.
ISBN 978-1-57864-647-0
1. Virginia Living Museum (Newport News, Va.) 2. Habitat conservation--Virginia--Newport News. 3. Natural history museums--Virginia--Newport News. 4. Museums--Virginia--Newport News. I. Hayhurst, Page.
QH76.5.V8V58 2010
591.9755--dc22
2010040571
The paper used to print this book is manufactured from wood grown in sustainable forests.

Contents

Museum entrance featuring bronze sculpture by Glenna Goodacre

Welcome

From the majestic Appalachian Mountains to the coast of the Chesapeake Bay and Atlantic Ocean, the Commonwealth of Virginia is blessed with an uncommon wealth of natural wonders. At the Virginia Living Museum this natural heritage comes alive in exhibits and programs for all ages.

Founded in 1966 by the Warwick Rotary Club and the Junior League of Hampton Roads, the Museum was first named the Peninsula Junior Nature Museum and Planetarium. In 1987, the Museum became the first living museum east of the Mississippi, combining the elements of a native wildlife park, science museum, aquarium, botanical garden and planetarium. Today, the Virginia Living Museum continues to be a national leader in connecting people to nature through educational experiences that promote conservation.

Visitors to the Museum encounter more habitats, wildlife and plant species than in a lifetime of outdoor adventures in Virginia. The exhibits showcase all of the state's regions and feature more than 250 different animal species. The Museum provides a critical sanctuary for injured and imprinted animals, most of which are non-releasable. The 21 acres of native plantings provide natural wildlife habitats for the exhibit animals and highlight earth-friendly tips for gardening and home construction.

In the Abbitt Planetarium visitors explore the universe in a state-of-the-art immersive planetarium theater. The Abbitt Observatory, the region's only public observatory, includes a university-grade telescope for searching the skies. In the Space Gallery visitors captain their own journey through the solar system with a *Magic Planet* interactive sphere.

The Museum's unique exhibits are a vital component in our extensive education programming. More than 2.5 million students have visited the Museum, experiencing hands-on science that is memorable and fun! In addition to student programs, the Museum hosts a wide range of professional development courses for science teachers as well as public programs and nature safaris for families.

The Museum is fortunate to have a committed and talented staff and a dedicated volunteer corps. Their work with the public and behind-the-scenes ensures the highest quality of visitor experience and animal care. Members and visitors from around the world also help support the Museum's education and conservation mission. We welcome you to join the VLM family and be ***Wild About Nature!***

Coastal Plain

Virginia's Coastal Plain reaches from the fall line near Richmond to the Atlantic Ocean. Only 250 feet at its highest elevation, it includes wetlands, marshes and the nation's richest sea life nursery, the Chesapeake Bay. Fossils of the region include those of woolly mammoths, mastodons, giant sloths and 50-foot sharks, an indication of the varied climate of the region's ancient past. Densely populated, Virginia's modern-day Coastal Plain is a center of commerce, tourism and industry where the sustainable management of coastal resources is critical.

Coastal Plain exhibits range from the dramatic 30,000-gallon Noland Chesapeake Bay Aquarium, with large species like sharks and a sea turtle, to a 20-gallon exhibit displaying a small glass lizard, whose tail can break off into numerous moving pieces when attacked by a predator.

"The country is...rivers and brooks, all running most pleasantly into a fair Bay."

Captain John Smith, *A Description of Virginia*

Found throughout Virginia's coastal region, the **Blue Crab (above)** *is one of the most economically important species in the Chesapeake Bay, with an annual value of 45 million dollars.*

Male **Seahorses (opposite page)** *carry up to 300 eggs in a pouch "kangaroo style." For several weeks the fathers transfer oxygen and nutrients to the developing young. When released after a few weeks, the tiny babies are able to care for themselves. Seahorses are also rare in the animal kingdom for keeping one mate for life.*

Conservation Note: Seahorses

Seahorses are at risk due to both harvesting by collectors and to the dramatic decline in underwater bay grasses that provide them protective habitat. The Museum hopes that by displaying seahorses it will heighten awareness and engage public support of marine conservation, while providing the opportunity to improve our knowledge of seahorse biology. The Museum participates in captive breeding programs with other facilities to supply the majority of its seahorses, rather than collecting them from the wild.

Chesapeake Bay Aquarium

Victoria Triska

The largest estuary in the country, the Chesapeake Bay supports more than 3,800 species of plants, fish and other animals. It also provides shelter for nearly a million waterfowl, many of them endangered and protected species. The name is derived from "Chesepiooc," from the native Algonquins. Containing over 30,000 gallons of seawater, this 11-foot deep aquarium is home to a loggerhead sea turtle, sharks and numerous fish species.

Victoria Triska

Animal Confidential: Loggerhead

The loggerhead is a summertime visitor to Virginia's waters and is the most common sea turtle to venture into the Chesapeake Bay. It is one of the largest reptiles in the Bay region, often exceeding weights of 300 lbs. The name *loggerhead* comes from the species' remarkably large head. Unlike box turtles, the loggerhead and other sea turtles cannot pull their heads into their shells. They have a body built for an aquatic life, with limbs that resemble paddles. These flippers move and steer them through the water rapidly.

Loggerheads face many threats in the wild including entanglement in fishing nets, loss of nesting sites due to increased beach development, strikes by passing boat propellers and mistaking floating plastic bags and balloons as their favorite food *(jellyfish)*. The Museum's loggerhead is fed by hand by the Museum's aquarists!

The Chesapeake Bay is visited by over 10 different species of sharks such as the ***Sandbar Shark (above)****. Due to overfishing, the shark population has decreased by 50% within the last 30 years.*

When you shop for seafood, choose fish that are plentiful and caught or farmed in ways that protect the oceans. The Virginia Living Museum's Seafood Watch guide, created in partnership with the Monterey Bay Aquarium, tells you which types to buy – and which types to avoid in the Southeast.

How many flounder can you find?

The **Summer Flounder** exhibit ***(top)*** highlights the fishes' amazing adaptation of being able to change colors to blend into different environments of the sea floor. Originally from the Indian Pacific ocean region, the **Lionfish *(lower left)*** is a venomous invader to Virginia's coastal ocean, most likely released from the pet trade. When threatened, the **Porcupine fish *(lower right)*** rapidly inflates its body by gulping water to become a hard-to-swallow pincushion.

Cypress Swamp

Cypress swamps are named for the deciduous bald cypress tree, which has a very wide base and root-like knees that rise out of the water. Bald cypress trees can grow to 120 feet tall. Southern Virginia is home to the Great Dismal Swamp with over 112,000 acres of forested wetlands.

The Cypress Swamp exhibit ***(opposite page)*** is a two-story recreation of a vital and endangered wetland habitat. This impressive gallery features an entire community of plants and swamp creatures including massive cypress and tupelo tree trunks, an alligator and snapping turtle. The exhibit features many live plants but the cypress knees and some tree trunks are concrete fabrications, molded from actual trees in a Mississippi swamp!

> *"It is remarkable that toward the middle of the Dismal (Swamp) no beast or bird or even reptile can live . . . because 'tis so overgrown with thickets that the genial beams of sun can never penetrate them."*
>
> William Byrd , *The Secret History of the Line*

The ***Eastern Snapping Turtle (above)*** *is the largest freshwater turtle in Virginia, weighing up to 50 lbs. and growing to 18 inches in length. Although aggressive on land, snapping turtles rarely bite people when underwater.*

Karlton Rebenstorf

Animal Confidential: The American Alligator

Rare among reptiles, alligators exhibit parental care behavior. Mothers will pick up and lightly break open unhatched eggs! Being very gentle, the mothers will carry the baby alligators to the water in their mouths. The babies remain with their mothers for the first year.

The American Alligator is one of the country's largest reptiles – up to 18 feet in length and 500 lbs. in weight! Having come back from near extinction in the 1960s, alligators can now be found less than 100 miles southeast of the Virginia Living Museum.

Cypress Swamp Birds

Karlton Rebenstorf

Green-winged Teal

Hooded Merganser

Visitors to the Cypress Swamp can listen for bird calls and the tapping of the Red- bellied Woodpecker (above) on the exhibit's trees.

Touch Tank

Touch horseshoe crabs, spider crabs, hermit crabs, whelks and sea stars with the help of the Museum's highly trained volunteer staff. ***Can you pass the Touch Tank true or false quiz?***

1. The horseshoe crab tail is venomous.
2. Spider crabs decorate their shells for camouflage.
3. Hermit crabs grow their shells.
4. Conchs are another name for whelks.
5. Sea stars can regenerate arms.

1.) F 2.) T 3.) F 4.) F 5.) T

Karlton Rebenstorf

What's the Buzz?

Since their introduction in the early 1600s, honeybees have been an essential part of agriculture in America as pollinators for numerous crops. Honeybees are social insects with wild hives containing up to 20,000 bees during the summer months. The bees work together in a highly structured social order. Each bee belongs to one of three castes: queen, drone and worker. There is only one queen per hive and she can lay over 1,500 eggs per day. The drones are males with the sole function of mating with a new queen. Worker bees do all the different tasks needed to maintain and operate the hive, including foraging outside the hive to gather nectar, pollen, and water.

The Museum's beehive exhibit gives visitors an insider's look into a working honey bee hive. The bees use a plexiglass tube from outdoors to gain access to their see-through hive area. Visitors can view within the hive the famous bee dance, the unique language of movements that tells other bees where to find food.

Piedmont and Mountains

Interstate highway I-95 is a rough guide to the location of a geologic boundary known as the fall line, where the Piedmont region begins. East of I-95, the Coastal Plain soil is sandy, light-colored and relatively flat. West of the interstate, the plowed fields expose red clay and the land rises steadily to the Blue Ridge and Appalachian Mountains.

Older than the Rockies and thought to have once been higher than the Himalayas, Virginia's Appalachian Mountains have become gently sculpted over the last 250 million years. The rich forests, streams and cool waterfalls are home to hundreds of unique animals and plants.

The Museum's exhibits include the fall line of the James River in Richmond ***(right)***, filled with smallmouth bass, sturgeon and wood turtles. Exhibits such as spotted turtles in an upland bog and chipmunks in their rock wall den help visitors piece together the connections between habitat, adaptation and survival.

Wendy Bardrof

*The **Common Mudpuppy** is a medium sized aquatic salamander found in clear lakes and streams, perfect habitat for many of Virginia's amphibians that need moist habitats to keep their skin from drying out. Yes, these salamanders can regenerate portions of their tail, and even entire limbs!*

Conservation Note: Shortnose Sturgeon

Shortnose sturgeon ***(above)***, seen in the James River exhibit, are one of the oldest fish in Virginia and one of the rarest. Swimming in our rivers before even the dinosaurs roamed, the sturgeon is one of the most primitive bony fish found in Virginia and has remained virtually unchanged for over 200 million years!

The shortnose sturgeon is Federally Endangered due to overfishing, pollution and dams that prevent spawning migrations. The Virginia Living Museum is the only facility in Virginia to exhibit shortnose sturgeon in collaboration with the National Marine Fisheries Service and U.S. Geological Survey.

Animal Confidential: Chipmunk and Spotted Turtle

The chipmunk's name comes from the *chip-chip-chip* sound that it vocalizes when alarmed. Chipmunks live in underground burrows with several hidden entrances and containing many rooms for storage, sleeping, a dump and a latrine. Some burrows even have a pantry that can store up to half a bushel of nuts and other food. Museum visitors have enjoyed watching chipmunk babies grow in the burrow!

The spotted turtle is one of the smaller turtles found in Virginia. Less than five inches long, spotted turtles are easily distinguished by their black shells with yellow or orange spots. Though the babies may have very few spots, they become more patterned and colorful as they age. Threatened by over-collection for the pet trade, spotted turtles are now protected by laws making it illegal to collect or possess them without a special state permit. Habitat loss, wetland degradation and pollution have also seriously impacted spotted turtle populations.

The cool, moist Appalachian Cove exhibit features a waterfall, a swift-running mountain stream, a lake filled with mountain fish and free-flying birds.

Stepping into this mountain cove exhibit provides the sights, sounds and experiences of an instant vacation to the mountains.

The **Rough Green Snake** ***(top)*** is well camouflaged in the foliage as it hunts for small insects, spiders and caterpillars.

Named for their distinctive call, **Northern Bobwhites** ***(center)*** roost on the ground in tight circles with their heads facing out.

In addition to the native brook trout, Virginia is now also home to brown and **Rainbow Trout** ***(bottom)***. They live in cool mountain streams and populations are highly regulated to protect existing stocks and to provide surplus for recreational fishing. At the Museum, the brook trout successfully spawn and their offspring are also on exhibit!

> *"Climb the mountains and get their good tidings. Nature's peace will flow into you as sunshine flows into trees."*
>
> John Muir (1901)

Wendy Bardrof

Karlton Rebenstorf

Karlton Rebenstorf

Wendy Bardrof

Conservation Notes: Paddlefish and Hellbender

Paddlefish *(left)* named for their paddle-like snout, are rare fish found in rivers in southwestern Virginia. Once believed to be used to dig prey out of the bottom, their elongate snout is actually an electrical antenna to help find food. Overfishing for its prized meat and eggs *(caviar)* has taken its toll on the paddlefish population. Paddlefish are a Federal Species of Concern as well as a state threatened species. As filter feeders, paddlefish are a challenge for any public aquarium where water must be continuously filtered. The Museum's paddlefish are trained to take small shrimp and fish directly from the hands of aquarium staff.

The **Eastern Hellbender** ***(above)***, the largest salamander in North America, can be found in Virginia's clear, fast flowing mountain streams. Hiding beneath stream rocks by day, hellbenders are rarely seen as they forage for food at night. They are also listed as a Federal Species of Concern and can only survive in stream systems isolated from development and urbanization. The Museum's hellbender is the most senior animal on exhibit at 25 years old!

Be the Box Turtle

Visitors of all ages enjoy climbing in and on the oversized box turtle located in one of the Museum's four different interactive Discovery Centers. These hands-on centers are brimming with specimens from the worlds of life, earth and space science.

Karlton Rebenstorf

Prehistoric Past

No dinosaur bones have ever been found in Virginia but they did leave many of their footprint trackways. The Museum exhibits a trackway left nearly 210 million years ago in what was then a vast swampy plain near Culpeper, Virginia. The nearby dinosaur sculpture is an artistic interpretation of a dynamic Syntarsus in action – paleontologists' best guess of the dinosaur that left these footprints.

For a wedding in the Museum, one family decorated the dinosaur sculpture with a bridal veil and named her *Bridezilla!*

Virginia Underground

Journey back through Virginia's ancient past by learning about fossil remains preserved beneath our feet and marvel at the beauty of rocks and minerals that formed deep under the Earth. Visitors can explore the colorful formations and critters of a limestone cave, wind their way past a fossil-rich river bank and peer into a jewel box tunnel of an underground gemstone mine.

In the Museum's fossil bank exhibit ***(left)***, visitors can compare fossils to similar creatures alive today to better visualize what life was like in ancient times. The Museum's fossil bank is a cross-section of the steep river banks along the James River near Chippokes State Park. The exhibit starts at the top with today's topsoil and goes down through the fossil-poor sediment from the ice ages to a fossil-rich layer from 3.5 to 5 million years ago when the ocean extended west to Richmond. The coral reef exhibit ***(below)*** represents what this part of the state looked like 3.5 to 5 million years ago based on this riverbank fossil evidence.

***Pack Rats** make their burrows at the entrances of caves, often using objects left behind by humans.*

Conservation Note

Caves are so important to their inhabitants and surrounding ecosystems that they are protected by state and federal laws. Water traveling through Virginia's caves eventually enters groundwater wells, springs and aquifers that supply drinking water. Caves are also home to ecologically important and rare species that could not survive anywhere else. Museum safaris take visitors and students to explore some of Virginia's 4,000 caves with expert naturalist guides.

World of Darkness

When we go to sleep at night another world awakens. For nocturnal animals, their survival depends on their sharpened senses to find food and escape enemies in the darkness. In the Museum's World of Darkness, visitors come eye-to-eye with tiny sharks, bats, scurrying ghost crabs, playful flying squirrels, eerie moon jellyfish and other animals that have unique adaptations for nightlife. These nocturnal animals are active during the day, tricked by the exhibits' reversed cycle of darkness in the day and light at night.

Described as cats with wings, **Eastern Screech Owls *(right)*** will attack animals much larger than themselves to defend a nest. Amazing hunters, owls have the best night vision of any animal and can hear a mouse squeaking from 75 feet away.

"For humans there is no wilderness more difficult to explore than the night."

David R. Wallace, *The Dark Range*

Wendy Bardrof

The **Barking Treefrog** *(above)* is the largest treefrog in the state, reaching lengths of over 2.5 inches. Like tiny gymnasts, treefrogs have long flexible legs to provide powerful jumps – up to 12 times their body length! The barking treefrog is listed as a state threatened species.

Using skin flaps between their wrists and ankles, **Southern Flying Squirrels** *(above)* launch themselves into daring nighttime glides of up to 200 feet. Their large dark eyes provide excellent night vision. The Museum's flying squirrels can often be seen resting together in a tree hollow.

Wendy Bardrof

Animal Confidential: Jellyfish and Lobster

Moon Jellyfish ***(above)*** rank among the most fascinating and beautiful creatures of the sea. They are made up of 95% water and are found in all oceans and in some freshwater habitats. The moon jellyfish is found mostly in warm coastal waters and ranges in size from 2 inches to several feet. Distinguished by its large bell, the moon jellyfish has small threadlike tentacles around the edge. Unlike the stinging nettle's painful wounds on area swimmers, the moon jellyfish's sting is not powerful enough to pierce through human skin. At the Museum, jellyfish are fed live brine shrimp, which are hatched daily by the aquarists.

The **American Lobster** ***(top right)*** reaches its southernmost habitat in the deep Norfolk Canyon 70 miles off Virginia's shoreline. Lobsters were once so common here that they were considered cheap food fed to servants and prisoners and even used for bait and fertilizer! While our lobster is large *(2.5 lbs.)*, the convex shape of the aquarium glass makes it look even bigger!

Wendy Bardrof

Stingrays are graceful swimmers, gliding through the water on "wings."

Outdoor Trail

The Museum's mile-long elevated boardwalk crosses Deer Park Lake and winds through woods and a wetlands area, highlighting animals native to Virginia in naturalized habitats. Interpretive stations along the boardwalk help explain the critical importance of Bay wetlands, showcase diverse pond life and challenge visitors to "ID a tree."

Karlton Rebenstorf

The 5,500 square-foot Coastal Plain Aviary is a dramatic walk-through aviary filled with coastal birds. In addition to the animals on display, visitors on the boardwalk can see fish in the stream, turtles on logs, bullfrogs along the streambanks and many different kinds of birds. Wild wood ducks, herons and other coastal birds actually try to get into the Museum's exhibits where food and shelter are plentiful!

Karlton Rebenstorf

Animal Confidential: Beaver

Beavers are Virginia's largest native aquatic rodent at 3 to 4 feet long and weighing up to 86 lbs. Beavers have webbed hind feet and large flat tails that act as a rudder as they swim. When alarmed, they will give a very loud slap with their tail to warn of danger.

Beavers live in dome-shaped lodges with one or more underwater entrances. They are herbivores, using their sharp incisor teeth to cut down vegetation for food and building material. Beavers will continue to rebuild dams and lodges that have been destroyed. These dams and lodges create wetland habitats while reducing downstream erosion and flooding.

The Museum's beavers can be seen building with twigs and branches placed in their outdoor habitat.

River Otter

River otters are remarkable and graceful swimmers. They can remain submerged for several minutes, dive to 55 feet and swim up to a quarter of a mile underwater. Special valves in their ears and nostrils keep water out. Dense, oily fur and a layer of fat under the skin keep otters warm even in icy water.

Otters are one of the most popular exhibits with viewing from a second story observation platform and under the water. Having been raised as a pet, the current male otter often plays to the crowd, diving and splashing. Otters are one of the few animals that play just for fun, often making slides on the river bank, similar to the stream slide built into the Museum's otter habitat!

Photos by Karlton Rebenstorf

Karlton Rebenstorf

Raccoon

The raccoon is one of the few examples of an animal that has thrived despite human development and sprawl. One reason raccoons are so successful is that they are true omnivores, eating a wide variety of foods from fruits and crayfish to eggs and even small mammals.

Raccoons often appear to be washing their hands when they are at the water's edge, although they are actually just naturally foraging and manipulating possible food items to better inspect them with their sensitive paws.

Photos by Karlton Rebenstorf

*The **Brown Pelican's** pouch attached to the beak can hold 2.5 gallons of water!*

Aviary

Virginia's Coastal Plain is located along the Atlantic Flyway, which channels the annual seasonal flight of millions of migratory birds. The region has always been a favored winter residence and stopover for many waterfowl species on their way to and from their northern breeding grounds. The wetlands, fields, shallows and open waters of the Bay offer a fertile environment for waterfowl to feed and rest.

In the aviary, visitors will find diving birds such as cormorants and canvasbacks, dabbling ducks *(those that feed "bottoms up")* such as pintails and wading birds like **Egrets *(right)*** and herons. The Museum provides supplemental heat in winter to make the migrating birds more comfortable!

Conservation Note: Red Wolves

In colonial times, red wolves ranged throughout the Southeast. Today they are the most endangered mammal in North America. The Museum's exhibit is the first public display of red wolves in Virginia. Two litters of pups have been born at the Museum as part of the federal Red Wolf Recovery captive breeding program. The program includes 180 red wolves at 38 zoos and museums around the country.

Red wolves are being reintroduced into the wild to prevent extinction of the species and to help restore ecosystems. Predators maintain the balance and health of ecosystems by controlling populations of prey species. As of 2010, there are approximately 100 red wolves divided into 20 packs or family groups across a designated 1.5 million-acre, five-county region that includes three national wildlife refuges. This national refuge area, only 100 miles southeast of the Museum, contains the only wild red wolves in the country. The biggest threat to the wild population is breeding with coyotes.

The Museum's pups weighed about one pound at birth, but within half a year they grew almost as big as their parents. The red wolf gets its name from the reddish color around its head, ears and legs.

Karlton Rebenstorf

Bald Eagle

The bald eagle, the national symbol of the United States, has a great recovery success story! In the late 1930s, the raptor's population began to shrink as a result of poaching, habitat loss and contamination from the pesticide DDT, which was used to control insects.

It wasn't until the 1960s that researchers found that DDT caused bald eagles and other birds to lay eggs with brittle shells that would crack when the parents sat on them.

In 1973 the bald eagle was listed as a Federally Endangered Species. With the ban of DDT and improved habitat protection, populations began recovering. In 2007 bald eagles were removed from the endangered species list!

The bald eagles on exhibit have wing injuries that make them unable to fly and nonreleasable.

Karlton Rebenstorf

JoAnne Rawls

Vultures

Visitors with weak stomachs may not want to learn more about the habits of the vulture. These birds dine on dead flesh, urinate on their legs to keep cool and regurgitate digested food before flight. Despite their poor table manners, vultures play an important role in the environment by removing animal carcasses from our forests, fields and highways.

The Museum displays both vulture species found in Virginia: the **Turkey Vulture** ***(left)*** and the **Black Vulture** ***(right)***. Both are non-releasable due to wing injuries.

Bobcat

One of the more secretive and fascinating mammals found in Virginia is the bobcat. Bobcats are adaptable, predatory members of the North American cat family. The only wild feline still found in the state, they are common in the Dismal Swamp and the mountains of Virginia but are rarely seen in other areas. Named for its short, bobbed, stubby tail, the bobcat is smaller than its northern cousin the lynx and twice as big as a large domestic cat.

Visitors often wonder if the bobcat on display gets lonely. In fact, these cats are usually solitary animals, except when a female is with her kittens or during mating season.

Photos by Karl Rebenstorf

White Tailed Deer

White tailed deer ***(shown on previous page)*** are graceful animals that are common throughout the eastern United States. With fewer predators, deer have now overpopulated many areas, resulting in less healthy herds. The tags seen on the Museum's deer identify them for veterinary inspection and treatment.

Wild Turkey

The wild turkey ***(right)*** is a large, ground-feeding bird that is related to chickens, quail and pheasants. Turkeys prefer to live in open forests, the edges of forests or wooded swamps. They feed on seeds, grains, nuts, berries and insects as well as an occasional frog or lizard. They are strong, quick flyers for short distances, flying up to 55 mph. Running turkeys have been clocked at 18 mph.

The Virginia Living Museum's turkeys are very active and can be seen foraging, displaying or taking dust baths.

Karlton Rebenstorf

Coyote

The Aztecs worshipped the coyote as a hunting god, a mischief-maker and a moon goddess for its baying at the moon. Native American legends portrayed it as a trickster: part god, part animal and part human. Americans are most familiar with its famous howl, which reminds us of the Old West.

Today coyotes are as much a part of the East as the West and can be found throughout Virginia. They thrive in both natural habitats and areas that have been disturbed by humans. Man's elimination of other large predators, such as wolves and mountain lions, has also made the coyotes' expansion possible. At the Museum, the coyotes and red wolves howl along with sirens from the nearby firehouse.

Photos by Karlton Rebenstorf.

Gray Fox

Red Fox

Sly as a Fox

Two members of this crafty canine family are found in Virginia, the red and gray fox. Red foxes were introduced to the United States from England in the mid-1800s. They prefer open habitats and sparsely settled hill areas.

Gray foxes prefer woodlands, brushy areas and swamplands. Both canines have adapted well to suburban environments and have been known to compete with raccoons for pet leftovers. Along with hunting and trapping, a new threat to foxes has been the expansion of the coyote population into Virginia. Coyotes compete successfully with foxes for food and have been known to attack their young.

The Museum's foxes enjoy enrichment activities, which simulate hunting behavior. *(See page 62 to learn more!)*

Photos by Karlton Rebenstorf.

Virginia Living Museum

Spring

Spring

Showers

Summer

Summer

Flowers

Through the Season*s*

Fall Leaves

Fall

Winter Snow

Winter

** Photos by Karlton Rebenstorf*

Conservation Gardens

The Virginia Living Museum's 3,000-square-foot Conservation Garden features earth-friendly gardening techniques: the use of native plants, mulching, composting, water conservation and reduced use of synthetic fertilizers and pesticides. It shows how proper landscaping methods can reduce storm water runoff that pollutes local waterways, while providing food, water and shelter for wildlife. The garden's decorative gate (above) was fabricated from 16th-inch steel by Northrop Grumman Newport News Shipbuilding.

A cold snap created this beautiful ice sculpture in the garden's walk-through waterfall.

Living Green

You've read all those articles about living green by maintaining your home in a manner that uses less energy, produces less waste and is kinder to Mother Earth. Visitors to the Museum's Goodson House can see all the latest techniques and products they can use to build and maintain an earth friendly home. The Goodson House is the first of its kind in Virginia and one of the first anywhere in the United States.

The Goodson House includes:

- Roofs covered with living plants, which provide insulation and reduce storm water runoff;
- Solar photovoltaic panels and a solar water heater;
- Alternative wall systems and insulation systems;
- Geothermal heat pump and cooling unit;
- Computer kiosk where visitors can calculate their own footprint on the earth's limited resources.

And more!

Virginia Garden

This historic garden highlights Virginia's botanical history from 1607 to the present. It includes the native plants of 1607 Virginia, the flora introduced to the Jamestown colonists by the Powhatan Indians and the plants brought to America from the colonists' European gardens. On display are many of the native trees and shrubs that early colonists sent back to Europe as exotic species.

The garden also showcases some non-native plants to avoid as they have become invasive and threaten native plant populations. Many of the vegetables and fruits grown in the garden help feed the Museum's animals!

> *"About their (Powhatan Indian) houses they have commonly square plots of cleared ground which serve them for gardens, some 100, some 200 foot square, wherein they sow their tobacco, pumpions, (...) guords and suchlike."*
>
> William Strachey (1612)

Karlton Rebenstorf

Ruby-throated Hummingbird

Tobacco Leaves

Fig Tree

French Lavender

Cantaloupe

Virginia Bluebe

The Museum's Holt Greenhouse ***(left)*** enables staff to propagate plants for exhibits and public native plant sales.

Karlton Rebenstorf

Butterfly Garden

The butterfly garden features a variety of native plant species that provide food and habitat for native butterflies and caterpillars. It includes both flowering plants whose nectar nourishes adult butterflies and specific host plants where butterflies lay their eggs. The leaves of these host plants provide a ready source of food for the caterpillars when they hatch.

In the fall, monarch butterfly chrysalides are collected from the Museum grounds in support of Monarch Watch, a cooperative network of researchers and citizen-scientists that track Monarch migration. After emerging from their chrysalides, adult **Monarch Butterflies** ***(left)*** are tagged and released for their long journey to Mexico. Several butterflies tagged by the Museum have been located along the migration route, including one found in Mexico and another found by a six-year-old boy in Austin, Texas – it traveled 1,306 miles in the 22 days since it had been released at the Museum!

Abbitt Planetarium

Magic Planet in the Space Gallery

In the Abbitt Planetarium ***(left)***, visit the planets, travel to a black hole, observe spinning galaxies or study the Earth from space. Using a digital full-screen projection system, planetarium staff can recreate the many wonders of the universe, including the stars and constellations of the sky for any season of the year – all for visitors to see from the comfort of an armchair.

Abbitt Observatory

With a dome that revolves 360 degrees and a university-grade, 16-inch Meade telescope, the rooftop Abbitt Observatory provides visitors with spectacular views of the universe at night. During the day, visitors can view sunspots, flares and solar prominences with the Museum's wide variety of portable telescopes.

Conservation Note

Where are all the stars? Excessive nighttime lighting in urban areas prevents us from seeing many stars and has negative effects on wildlife *(especially baby sea turtles and migrating night birds)*. The Virginia Living Museum is improving its lighting practices to prevent unneeded light from shining up into the sky. Please support this quest to bring back the night!

Behind-the-Scenes

Science Education

Science comes alive for students of all ages at the Virginia Living Museum. On a spring day, the Museum's 4 classrooms, science lab, planetarium, pre-K center, canoe rigs and floating dock may all be in use at once by students of all ages!

Middle school students engage in ecology field work about pond life on the Museum's floating dock ***(top left)***. Preschool students learn how plants grow in the Children's Learning Garden ***(center).*** Outdoor education such as canoe programs ***(right)*** promote healthy living as well as experiential learning.

Museum educators also provide nationally recognized teacher training programs which are offered for college credit through a partnership with Hampton University.

The Museum maintains a collection of over 48,000 specimens ranging from animal mounts and bird nests to meteorites and fossils. Museum specimens are used in exhibits and programs, as well as in discovery boxes that send hands-on science materials to students throughout the state.

Karlton Rebenstorf

Not all of the Museum's animals are in permanent exhibits. Many are specially trained as program animals, acting as animal ambassadors to help people connect to nature. The Museum's program animals can be seen daily in shows at the Wason Amphitheater during the summer as well as in travel programs to schools, day care centers and libraries.

The first question asked about program animals is *"What's its name?"* Visitors are surprised to learn that we choose not to name our animals to emphasize that these are wild creatures, rather than pets and that they should be observed rather than *disturbed* when encountered in the wild.

Red-tailed Hawk

Karlton Rebenstorf

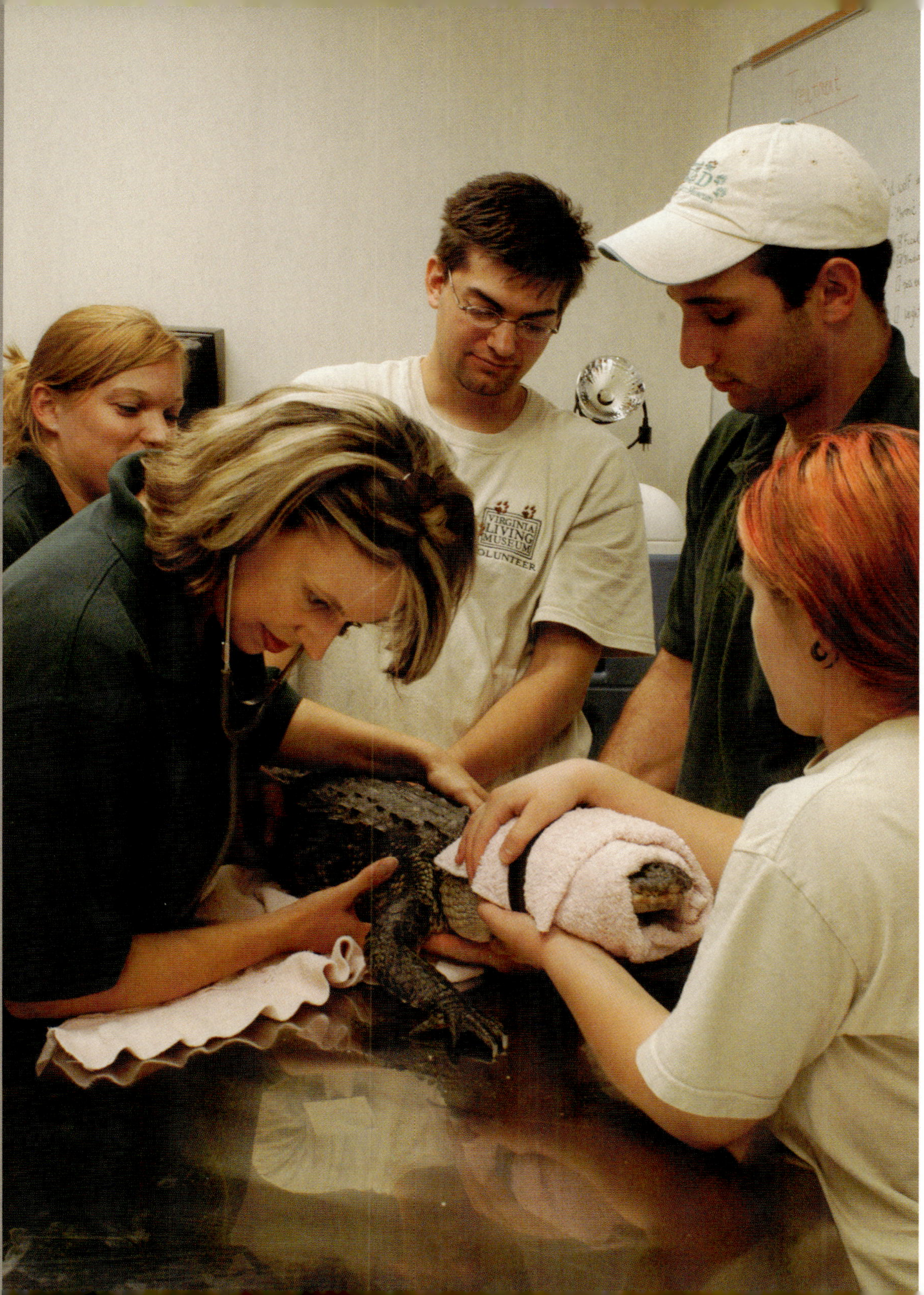

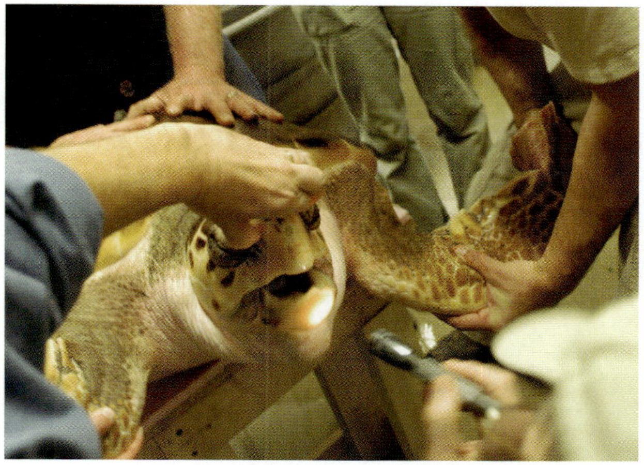

Animal Care

Veterinary care at the Virginia Living Museum includes a team of veterinarians with different specialties as well as in-house care from our veterinary technician. With support from this skilled vet team, the animal keepers optimize the health and welfare of the Museum's diverse collection of animals.

The Museum's animal care facilities include radiology, surgery, treatment and examination areas, a diet preparation kitchen and separate quarantine facilities. Quarantine facilities are used for all new animals to prevent disease introduction as well as for exhibit animals under medical treatment.

Disease prevention is a key role of the veterinary department. Animals are given healthy diets, regularly scheduled health check-ups and given needed vaccinations. Since many of the Museum's animals live longer in captivity than in the wild, they often experience the same health problems as pets and people do with aging *(like arthritis)*!

VLM Grocery List

28 lbs. Vegetables (sweet potatoes, carrots, broccoli)
25 lbs. Fruit (bananas, grapes, assorted berries)
5,000 Mealworms
200 Night Crawlers
2,000 Live Crickets
60 lbs. Meat
140 Mice
43 lbs. Squid and Shellfish
210 lbs. Fish (croaker and herring)
250 Live Minnows / Shiners
9 lbs. Gel Diet (fish/reptile)
150 lbs. Animal Chow (dog chow, turkey chow, and turtle pellets)

Above: *Not your typical shopping list! Each week the Museum requires 650 lbs. of food to feed the animals including some unusual items and quantities not found on most people's grocery list…*

The groundhog enjoys a snack of broccoli.

Karlton Rebenstorf

Daily Specials!

Volunteers help prepare a variety of diets as shown on this diet board ***(bottom right).*** Feeding hundreds of animals each day can be quite a challenge! Animal diets are carefully researched and lab tested to ensure they are providing the nutrients the Museum's animals need to thrive. Many of the animals must be hand fed by keepers.

Aquarists must scuba dive in the Museum's larger tanks to clean exhibits and to provide enrichment activities.

How Does the Museum Acquire Animals?

Many of our animals are received as non-releasable animals from rehabilitators and wildlife centers that take in injured native animals. The Museum also acquires animals from other zoos and aquariums, sometimes in special management programs such as the Red Wolf Species Survival Program. Some animals are bred at the Museum, such as mice, chipmunks, trout, night herons, black snakes and cottonmouths. Some local fish species are collected in the wild by our aquarists and are displayed in different tanks as they grow from a juvenile to an adult size!

Animals such as orphaned baby opossums are often raised by hand at the Museum by animal handlers to become either exhibit or program animals.

Enrichment

Karlton Rebenstorf

Unlike their wild counterparts, animals at the Museum do not have to forage for food, find shelter or escape predators. Since these natural behaviors keep wild animals fit and focused, animal keepers create enrichment activities to maintain optimal health of the animals.

Specific enrichment programs for each animal occur in the Animal Playground area on the outdoor trail, behind-the-scenes or even on exhibit. Activities include hidden foods, special scents and unusual objects added to the habitats. In addition to stimulating natural behaviors, training is also used to make it less stressful for animals to receive medication.

The ***Red Wolf (left)*** *enjoys a special chew treat. The* ***Striped Skunk (above)*** *works a puzzle feeder looking for mealworms.*

Photos by Karlton Rebenstorf

Unusual objects can trigger all kinds of natural behaviors from play to aggression. In the photo above, a stuffed toy brought out the fox's hunting instincts.

A game with a reward helps provide mental stimulation. For example, the Museum's raccoons ***(top left)*** have been trained by staff to paint!

Enrichment may include special food treats with new textures or flavors – in the summer, blocks of ice with frozen treats inside are a favorite! River Otters ***(bottom left)*** enjoy playing with their food!

Virginia Living Museum

Wild about Nature!

Edited by
Museum Executive Director Page Hayhurst.

Layout by Design Assistant Kasha Lang

Copywriters included Page Hayhurst,
Deputy Director Fred Farris,
Education Director Christine Lewis,
Marketing Director Virginia Gabriele
and Curatorial Director George Mathews.

Additional Photos by Marketing Manager Judy Triska,
VLM Staff and Visitors